ANCIENT MACHINES
AND CONTRIVANCES

Dollarbook No. 2

Coronado Press   1972

Copyright © 1972
by
Coronado Press
Box 3232
Lawrence, Kansas 66044

Standard Book Number 87291–035–0

Manufactured in the USA

General Editor   John E. Longhurst

Preface

The material included herein is taken, with some minor modifications of the highly technical nature of the originals, from William Smith (ed.), *Dictionary of Greek and Roman Antiquities*, 2nd edition, London, 1856. The title of the book is borrowed from Dr. Smith's own subject-matter index, and includes all those items and illustrations listed by him under "Ancient Machines and Contrivances."

JEL

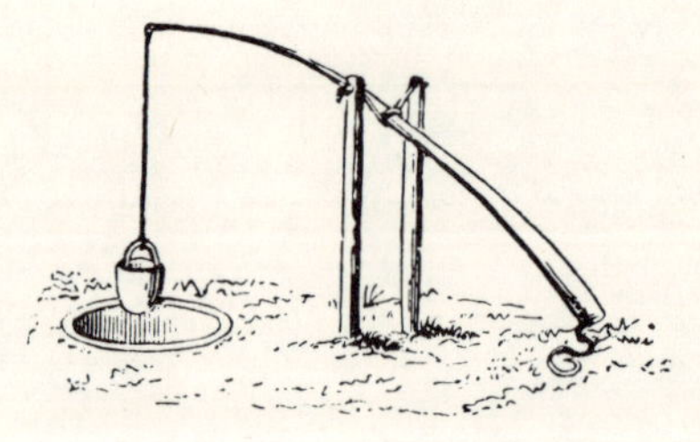

Antlia

ANTLIA (ἀντλία), any machine for raising water; a pump. The annexed figure shows a machine which is still used on the river Eissach in the Tyrol, the ancient Atagis. As the current puts the wheel in motion, the jars on its margin are successively immersed and filled with water. When they reach the top, the water is sent into a trough, from which it is conveyed to a distance, and chiefly used for irrigation.

Lucretius (v. 517) mentions a machine constructed on this principle:—"Ut fluvios versare rotas atque haustra videmus."

In situations where the water was at rest, as in a pond or a well, or where the current was too slow and feeble to put the machine in motion, it was constructed so as to be wrought by animal force, and slaves or criminals were commonly employed for the purpose (εἰς ἀντλίαν καταδικασθῆναι Artemid. *Oneiroc.* i. 50; *in antliam condemnare,* Suet. *Tib.* 51.) Five such machines are described by Vitruvius, in addition to that which has been already explained, and which, as he observes, was turned *sine operarum calcatura, ipsius fluminis impulsu.* These five were, 1. the tympanum; a tread-wheel, wrought *hominibus calcantibus*: 2. a wheel resembling that in the preceding figure; but having, instead of pots, wooden boxes or buckets (*modioli quadrati*), so arranged as to form steps for those who trod the wheel: 3. the chain-pump: 4. the *cochlea,* or Archimedes' screw: and 5. the *ctesibica machina,* or forcing-pump. (Vitruv. x. 4-7; Drieberg, *Pneum. Erfindungen der Griechen,* p. 44-50.)

On the other hand, the antlia with which Martial (ix. 19) watered his garden, was probably the pole and bucket universally employed in Italy, Greece, and Egypt. The pole is curved, as shown in the annexed figure; because it is the stem of a fir, or some other tapering tree. The bucket, being attached to the top of the tree, bends it by its weight; and the thickness of the other extremity serves as a counterpose. The great antiquity of this method of raising water is proved by representations of it in Egyptian paintings. (Wilkinson, *Manners and Cust. of Anc. Egypt,* ii. 1-4; see also *Pitt. d'Ercolano,* vol. i. p. 257.)

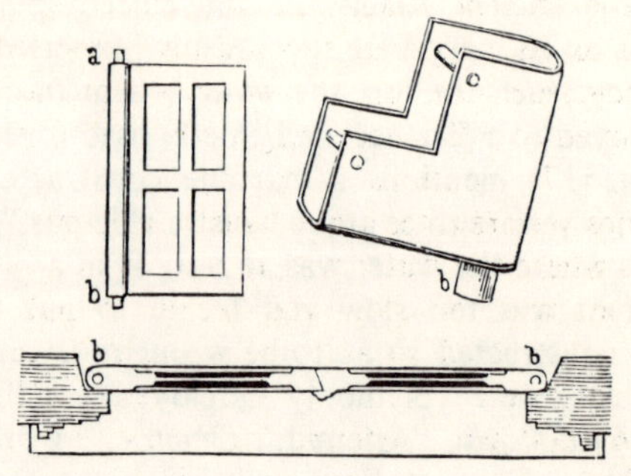
a
b
b

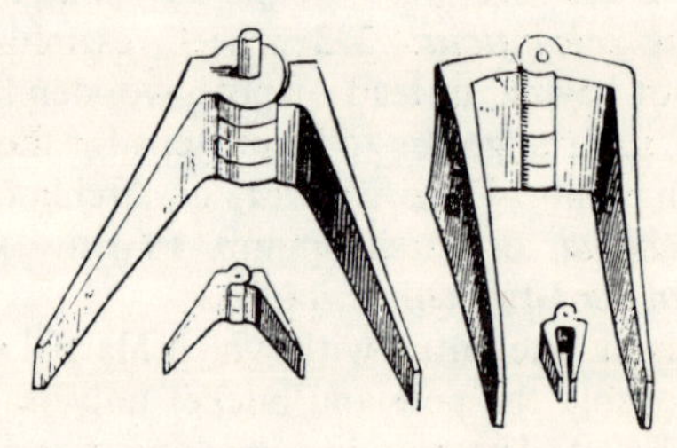
b
b

Cardo

CARDO (δαιρός, στροφεύς, στρόφιγξ, γίγγλυμος), a hinge, a pivot. The first figure in the annexed woodcut is designed to show the general form of a door, as we find it with a pivot at the top and bottom (*a,b*) in ancient remains of stone, marble, wood, and bronze. The second figure represents a bronze hinge in the Egyptian collection of the British Museum: its pivot (*b*) is exactly cylindrical.

Under these is drawn the threshold of a temple, or other large edifice, with the plan of the folding doors.

The pivots move in holes fitted to receive them (*b,b*), each of which is in an angle behind the antepagmentum (*marmoreo aeratus stridens in limine cardo*, Virg. *Ciris,* 222; Eurip. *Phoen.* 114-116, Schol. *ad loc.*).

The Greeks and Romans also used hinges exactly like those now in common use. Four Roman hinges of bronze, preserved in the British Museum, are here shown.

The form of the door above delineated makes it manifest why the principal line laid down in surveying land was called "cardo" (Festus, *s.v. Decumanus*; Isid. *Orig.* xv. 14); and it further explains the application of the same term to the North Pole, the supposed pivot on which the heavens revolved. (Varr. *De Re Rust.* i.2; Ovid, *Ex Ponto,* ii.10.45.)

The lower extremity of the universe was conceived to turn upon another pivot, corresponding to that at the bottom of the door (Cic. *De Nat. Deor.* ii.41; Vitruv. vi.1, ix.1); and the conception of these two principal points in geography and astronomy led to the application of the same term to the East and West also. (Lucan. v.71.) Hence our "four points of the compass" are called by ancient writers *quatuor cardines orbis terrarum*, and the four principal winds, N.S.E. and W., are the *cardinales venti*. (Serv. *ad Aen.* i.85.)

Catena

CATENA, dim. CATELLA (ἄλυσις, dim. ἀλύσιον, ἀλυσίδιον), a chain. The chains which were of superior value, either on account of the material or the workmanship, are commonly called *catellae* (ἀλύσια), the diminutive expressing their fineness and delicacy as well as their minuteness.

The specimens of ancient chains which we have in bronze lamps, in scales, and in ornaments for the person, especially necklaces, show a great variety of elegant and ingenious patterns. Besides a plain circle or oval, the separate link is often shaped like the figure 8, or is a bar with a circle at each end, or assumes other forms, some of which are here shown. The links are also found so closely entwined, that the chain resembles platted wire or thread, like the gold chains now manufactured at Venice. This is represented in the lowest figure of the woodcut.

These valuable chains were sometimes given as rewards to the soldiers (Liv. xxxiv.31); but they were commonly worn by women (Hor. *Ep.* i.17.55), either on the neck (περὶ τὸν τράχηλον ἀλύσιον, Menander, p. 92, ed. Mein.), or round the waist (Plin. *H.N.* xxxiii.12); and were used to suspend pearls, or jewels set in gold, keys, lockets, and other trinkets.

[11]

Clitellae

CLITELLAE, a pair of panniers, and therefore only used in the plural number. (Hor. *Sat.* i.5.47; Plaut. *Most.* iii.2.91.) In Italy they were commonly used with mules or asses, but in other countries they were also applied to horses, of which an instance is given in the annexed woodcut from the column of Trajan; and Plautus (*Ib.* 94) figuratively describes a man upon whose shoulders a load of any kind, either moral or physical, is charged, as *homo clitellarius.*

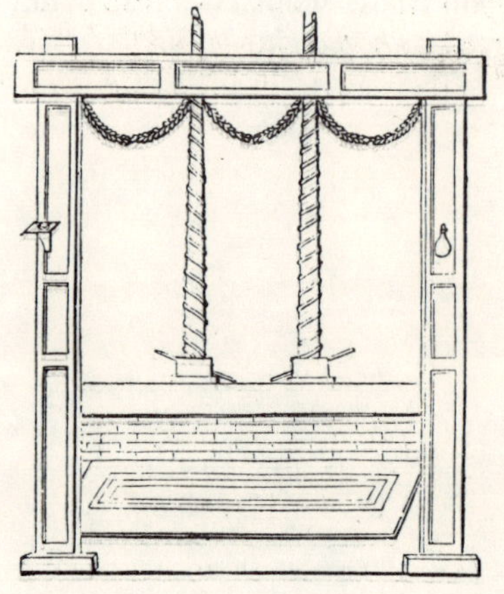

Cochlea

COCHLEA (κοχλίας), which properly means a snail, was also used to signify other things of a spiral form.

1. A screw. The woodcut annexed represents a clothes-press, from a painting on the wall of the Chalcidicum of Eumachia, at Pompeii, which is worked by two upright screws (*cochleae*) precisely in the same manner as our own linen presses. (*Mus. Borbonico,* iv.50.)

A screw of the same description was also used in oil and wine presses. (Vitruv. vi.9. p.180, ed. Bipont.; Palladius, iv.10. § 10, ii.10. § 1.) The thread of the screw, for which the Latin language has no appropriate term, is called περικόχλιον in Greek.

2. A spiral pump for raising water, invented by Archimedes (Diod. Sic. i.34, v.37; compare Strab. xvii.30), from whom it has ever since been called the Archimedean screw. It is described at length by Vitruvius (x.11).

3. A peculiar kind of door, through which the wild beasts passed from their dens into the arena of the amphitheatre. (Varr. *De Re Rust.* iii.5. § 3.) It consisted of a circular cage, open on one side like a lantern, which worked upon a pivot and within a shell, like the machines used in the convents and foundling hospitals of Italy, termed *rote,* so that any particular beast could be removed from its den into the arena merely by turning it round, and without the possibility of more than one escaping at the same time; and therefore it is recommended by Varro (*l.c.*) as peculiarly adapted for an aviary, so that the person could go in and out without affording the birds an opportunity of flying away. Schneider (in *Ind. Script. R.R. s. v. Cavea*), however, maintains that the *cochlea* in question was nothing more than a portcullis (*cataphracta*) raised by a screw, which interpretation does not appear so probable as the one given above.

Columbarium

COLUMBARIUM, literally a dove-cote or pigeon-house, is used to express a variety of objects, all of which however derive their name from their resemblance to a dove-cote.

In a machine used to raise water for the purpose of irrigation, as described by Vitruvius (x.9), the vents through which the water was conveyed into the receiving trough, were termed *Columbaria.* This will be understood by referring to the woodcut of **ANTLIA**, above. The difference between that representation and the machine now under consideration, consisted in the following points:—

The wheel of the latter is a solid one (*tympanum*), instead of radiated (*rota*); and was worked as a treadmill, by men who stood upon platforms projecting from the flat sides, instead of being turned by a stream.

Between the intervals of each platform a series of grooves or channels (*columbaria*) were formed in the sides of the tympanum, through which the water taken up by a number of scoops placed on the outer margin of the wheel, like the jars in the cut referred to, was conducted into a wooden trough below (*labrum ligneum suppositum,* Vitruv. *l.c.*).

Ephippium

EPHIPPIUM (ἀστράβη, ἐφίππιον, ἐφίππειον), a saddle. Although the Greeks occasionally rode without any saddle (ἐπὶ ψιλοῦ ἵππου, Xenoph. *De Re Eques.* vii.5), yet they commonly used one, and from them the name, together with the thing, was borrowed by the Romans. (Varr. *De Re Rust.* ii.7; Caes. *B. G.* iv.2; Hor. *Epist.* i.14.43; Gellius, v.5.)

It has indeed been asserted, that the use of saddles was unknown until the fourth century of our era. But Ginzrot, in his valuable work on the history of carriages (vol. ii. c.26), has shown, both from the general practice of the Egyptians and other Oriental nations, from the pictures preserved on the walls of houses at Herculaneum, and from the expressions employed by J. Caesar and other authors, that the term "ephippium" denoted not a mere horse-cloth, a skin, or a flexible covering of any kind, but a saddle-tree, or frame of wood, which, after being filled with a stuffing of wool or cloth, was covered with softer materials, and fastened by means of a girth (*cingulum, zona*) upon the back of the animal.

The ancient saddles appear, indeed, to have been thus far different from ours, that the cover stretched upon the hard frame was probably of stuffed or padded cloth rather than leather, and that the saddle was, as it were, a cushion fitted to the horse's back. Pendent cloths (στρώματα, *strata*) were always attached to it so as to cover the sides of the animal; but it was not provided with stirrups.

As a substitute for the use of stirrups the horses, more particularly in Spain, were taught to kneel at the word of command, when their riders wished to mount them. See the illustration from an antique lamp found at Herculaneum, and compare Strabo, iii.1. p.436, ed. Sieb.; and Silius Italicus, x.465.

The saddle with the pendent cloths is also exhibited in the annexed coin of Q. Labienus.

The term "Ephippium" was in later times in part supplanted by the word "sella" and the more specific expression "sella equestris."

Exostra

EXOSTRA (ἐξώστρα, from ἐβωθέω), was one of the many kinds of machines used in the theatres of the ancients. Cicero (*De Prov. Cons.* 6), in speaking of a man who formerly concealed his vices, expresses this sentiment by *post siparium heluabatur*; and then stating that he now shamelessly indulged in his vicious practices in public, says, *jam in exostra heluatur.* From an attentive consideration of this passage, it is evident that the exostra was a machine by means of which things which had been concealed behind the siparium, were pushed or rolled forward from behind it, and thus became visible to the spectators. This machine was therefore very much like the ἐκκύκλημα, with this distinction, that the latter was moved on wheels, while the exostra was pushed forward upon rollers. (Pollux, iv.128; Schol. ad Aristoph. *Acharn.* 375.) But both seem to have been used for the same purpose; namely, to exhibit to the eyes of the spectators the results or consequences of such things — *e.g.* murder or suicide — as could not consistently take place in the proscenium, and were therefore described as having occurred behind the siparium or in the scene.

The name exostra was also applied to a peculiar kind of bridge, which was thrown from a tower of the besiegers upon the walls of the besieged town, and across which the assailants marched to attack those of the besieged who were stationed on the ramparts to defend the town. (Veget. *De Re Milit.* iv.21.)

[20]

Ferculum

FERCULUM (from *fer-o*), is applied to any kind of tray or platform used for carrying anything. Thus it is used to signify the tray or frame on which several dishes were brought in at once at dinner (Petron. 35; Plin. *H.N.* xxviii.2); and hence *fercula* came to mean the number of courses at dinner, and even the dishes themselves. (Suet. *Aug.* 74; Serv. *ad Virg. Aen.* i.637; Juv. i.93, xi.64; Hor. *Sat.* ii.6.104; Mart. iii.50, ix.82, xi.31.)

The ferculum was also used for carrying the images of the gods in the procession of the circus (Suet. *Jul.* 76), the ashes of the dead in a funeral (Suet. *Cal.* 15), and the spoils in a triumph (Suet. *Jul.* 37; Liv. i.10); in all which cases it appears to have been carried on the shoulders or in the hands of men. The most illustrious captives were sometimes placed on a ferculum in a triumph, in order that they might be better seen. (Senec. *Herc. Oet.* 109.)

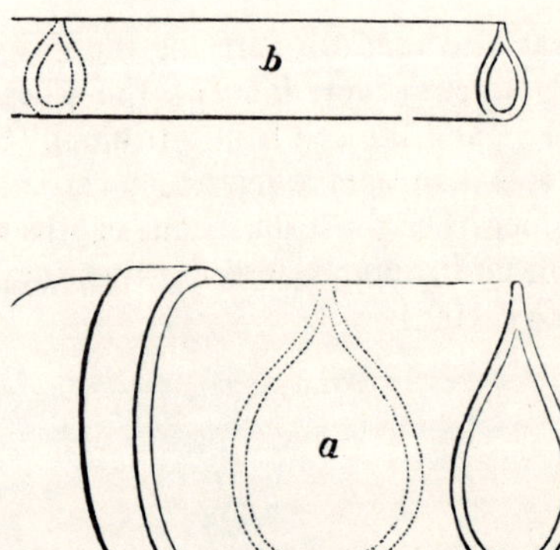

b
a

Fistula

FISTULA (σωλήν), a water-pipe. Vitruvius (viii.7. s.6. § 1, ed. Schn.) distinguishes three modes of conveying water: by channels of masonry (*per canales structiles*), by leaden pipes (*fistulis plumbeis*), and by earthen pipes (*tubulis fictilibus*); but of these two sorts of pipes the leaden were the more commonly used.

[The etymological distinction between *fistula* and *tubus* seems to be that the former, which originally signified a *flute,* was a small pipe, the latter a large one; but, in usage, at least so far as water-pipes are concerned, it seems that *fistula* is applied to a leaden pipe, *tubus* and *tubulus* to one of any other material, especially of terra-cotta.]

They were made by bending up cast plates of lead into a form not perfectly cylindrical, but having a sort of ridge at the junction of the edges of the plate, as represented in the following engraving, taken from antique specimens. (Frontin. *de Aquaed.* p.73. fig. 15, 16, ed. Polen.; Hirt, *Lehre d. Gebäude,* pl. xxxii. fig. 8.)

In the manufacture of these pipes, particular attention was paid to the bore, and to the thickness. The accounts of Vitruvius, Frontinus, and other writers, are not in perfect accordance; but it appears, from a comparison of them, that two different systems of measurement were adopted, namely, either by the width of the plate of lead (*lamina* or *lamna*) before it was bent into the shape of a pipe, or by the internal diameter or bore (*lumen*) of the pipe when formed.

The former is the system adopted by Vitruvius (*l.c.* § 4); according to him the leaden plates were cast of a length not less than ten feet, and of a width containing an exact number of *digits* (sixteenths of a foot), which number was of course different for different sized pipes; and then the sizes of the pipes were named from the number of digits in the width of the plates, as in the following table, where the numbers on the right hand indicate the number of pounds which Vitruvius assigns to each ten-feet length of pipe:—

Centenaria, from a plate	100 digits wide:	1200 lbs.
Octogenaria —	80 —	960 —
Quinquagenaria —	50 —	600 —
Quadragenaria —	40 —	480 —
Tricenaria —	30 —	360 —
Vicenaria —	20 —	240 —
Quindena —	15 —	180 —

Dena	—	10	—	120 –
Octona	—	8	—	96*–
Quinaria	—	5	—	60 –

*(*Pliny and Palladius, and even the ancient MSS. of Vitruvius, give here C, which, however, is clearly an error of a transcriber who did not perceive the law of the proportion, but who had a fancy for the round number.)*

From this scale it is evident, at a mere glance, that the thickness of the plates was the same for pipes of all sizes, namely, such that each strip of lead, ten feet long and one digit wide, weighed twelve pounds. The account of Vitruvius is followed by Pliny (*H.N.* xxxi.6 s.31) and Palladius (ix.12: comp. the notes of Schneider and Gesner).

Frontinus, who enters into the subject much more minutely, objects to the system of Vitruvius as too indefinite, on account of the variation which is made in the shape of the pipe in bending up the plate of lead; and he thinks it more probable that the names were derived from the length of the internal diameters, reckoned in *quadrantes* (the unit being the digit), that is, *in quarters of a digit*; so that the *Quinaria* had a diameter of five fourths of a digit, or 1¼ digit, and so on, up to the *Vicenaria,* above which the notation was altered, and the names were no longer taken from the number of *linear quarters of a digit* in the *diameter* of the pipe, but from the number of *square quarters of a digit* in its area, and this system prevailed up to the *Centumvicena,* which was the largest size in use, as the *Quinaria* was the smallest: the latter is adopted by Frontinus as the standard measure (*modulus*) of the whole system. (For further details see Frontinus, *de Aquaed.* 20-63, pp.70-112, with the Notes of Polenus.)

Another mode of explaining the nomenclature was by the story that when Agrippa undertook the oversight of the aquaeducts, finding the *modulus* inconveniently small, he enlarged it to *five* times its diameter, and hence the origin of the *fistula quinaria.* (Frontin. 25, pp. 80,81.) Of these accounts that of Vitruvius appears at once the most simple and the most correct: indeed it would seem that the plan of measurement was very probably the invention of Vitruvius himself. (Frontin. *l.c.*)

Of the earthen (terra-cotta) pipes we know very little. Pliny says that they are best when their thickness is two digits (1½ inch), and that each pipe should have its end inserted in the next, and the joints should be cemented; but that leaden pipes should be used where the water rises. The earthen pipes were thought more wholesome than the leaden. (Plin. *H.N.* xxxi.6. s.31; Vitruv. *l.c.* § 10; Pallad. ix.11.) Water pipes

[24]

were also made of leather (Plin. *H.N.* v.31. s.34; Vitruv. *l.c.* § 8); and of wood (Pallad. *l.c.*), especially of the hollowed trunks of the pine, fir, and alder. (Plin. *H.N.* xvi.42. s.81.)

Follis

FOLLIS, *dim.* FOLLICULUS, an inflated ball of leather, perhaps originally the skin of a quadruped filled with air: Martial (iv.19) calls it "light as a feather." Boys and old men among the Romans threw it from one to another with their arms and hands as a gentle exercise of the body, unattended with danger. (Mart. vii.31, xiv.45,47; Athen. i.25.) The emperor Augustus (Suet. *Aug.* 83) became fond of the exercise as he grew old. (See Becker, *Gallus,* vol.i. p.271.)

The term *follis* is also applied to a leather purse or bag (Plaut. *Aul.* ii.4.23; Juv. xiv.281); and the diminutive *folliculus* to the swollen capsule of a plant, the husk of a seed, or anything of similar appearance. (Senec. *Nat. Quaest.* v.18; Tertull. *De Res. Carn.* 52.)

Two inflated skins (δύο φύσαι, Herod. i.68; ζώπυρα, Ephor. *Frag.* p.188; πρηστῆρες, Apoll. Rhod. iv.763,777), constituting *a pair of bellows,* and having valves adjusted to the natural apertures at one part for admitting the air, and a pipe inserted into another part for its emission, were an essential piece of furniture in every forge and foundry. (*Il.* xviii.372-470; Virg. *Aen.* viii.449.)

According to the nature and extent of the work to be done the bellows were made of the hides of oxen (*taurinis follibus,* Virg. *Georg.* iv.171), or of goats (*hircinis,* Hor. *Sat.* i.4.19), and other smaller animals.

The nozzle of the bellows was called ἀκροφύσιον or ἀκροστόμιον (Thucyd. iv.100; Eust. in *Il.* xviii.470).

In bellows made after the fashion of those exhibited in the lamp here introduced from Bartoli (*Ant. Lucerne,* iii.21), we may imagine the skin to have been placed between the two boards so as to produce a machine like that which we now employ.

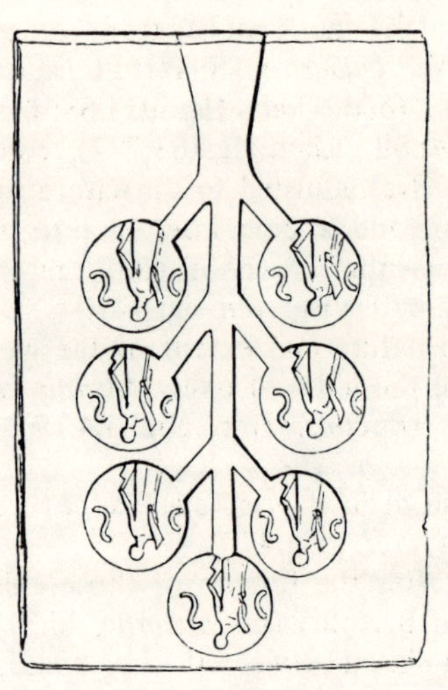

FORMA, *dim.* FORMULA, *second dim.* FORMELLA (τύπος), a pattern, a mould; any contrivance adapted to convey its own shape to some plastic or flexible material, including moulds for making pottery, pastry, cheese, bricks, and coins.

The moulds for coins were made of a kind of stone, which was indestructible by heat. (Plin. *H.N.* xxxvi.49.) The mode of pouring into them the melted metal for casting the coins will be best understood from the annexed woodcut, which represents one side of a mould, engraved by Seroux d'Agincourt.

Moulds were also employed in making walls of the kind, now called *pisé,* which were built in Africa, in Spain, and about Tarentum. (Varro, *De Re Rust.* i.14; Pallad. i.34; *parietes formacei,* Plin. *H.N.* xxxv.48.)

The shoemaker's last was also called *forma* (Hor. *Sat.* ii.3.106) and *tentipellium* (Festus, *s.v.*), in Greek καλόπους. (Plato, *Conviv.* p. 404, ed. Bekker.)

The spouts and channels of aqueducts are called *formae,* perhaps from their resemblance to some of the moulds included in the above enumeration. (Frontin. *De Aquaeduct.* 75,126.)

Fornax

FORNAX, *dim.* FORNACULA (κάμινος *dim.* καμίνιον), a kiln; a furnace. The construction of the kilns used for baking earthenware may be seen in the annexed woodcut, which represents part of a Roman pottery discovered at Castor, in Northamptonshire. (Artis's *Durobrivae,* Lond. 1828). The dome-shaped roof has been destroyed; but the flat circular floor on which the earthenware was set to be baked is preserved entire.

The middle of this floor is supported by a thick column of brick-work, which is encircled by the oven (*furnus,* κλίβανος). The entrance to the oven (*praefurnium*) is seen in front.

The lower part of a smelting-furnace, shaped like an inverted bell, and sunk into the earth, with an opening and a channel at the bottom for the discharge of the melted metal, has been discovered near Arles. (Florencourt, *über die Bergwerke der Alten,* p. 30.)

In Spain these furnaces were raised to a great height, in order that the noxious fumes might be carried off. (Strabo, iii.2. p. 391, ed. Sieb.) They were also provided with long flues (*longinquae fornacis cuniculo,* Plin. *H.N.* ix.62), and with chambers (*camerae*) for the purpose of collecting more plentifully the oxides and other matters by sublimation (*Ibid.* xxxiv. 22. 33–41).

Homer describes a blast-furnace with twenty crucibles (χοανοί, *Il.* xviii. 470).

Melting-pots or crucibles have been found at Castor (Artis, pl. 38), and at different places in Egypt, in form and material very like those which we now employ. (Wilkinson, *Man. and Cust.* vol.iii. p.224.)

A glass-house, or furnace for making glass, was called ὑελουργεῖον. (Dioscor. v.182.)

Furnaces of an appropriate construction were erected for casting large statues of bronze (Claud. *De Laud. Stil.* ii.176), and for making lamp-black. (Vitruv. vii.10.)

The lime-kiln (*fornax calcaria*) is described by Cato. (*De Re Rust.* 38; see also Plin. *H.N.* xvii.6; Vitruv. vii.3.)

The early Romans recognized, under the name of Fornax, a divinity who presided over ovens and furnaces.

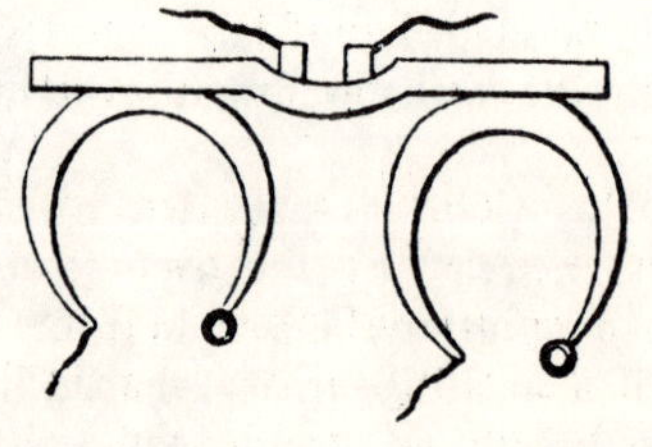

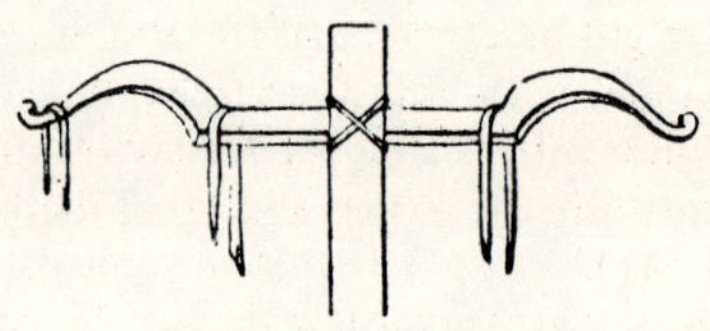

Jugum

JUGUM (ζυγὸς, ζυγὸν), the yoke by which ploughs and carriages were drawn. The yoke was in many cases a straight wooden plank or pole laid upon the horses' necks; but it was commonly bent towards each extremity, so as to be accommodated to the part of the animal which it touched.

The annexed woodcut shows two examples of the yoke, the upper from a MS. of Hesiod's *Works and Days,* preserved at Florence, the lower from a MS. of Terence belonging to the Vatican library.

The practice of having the yoke tied to the horns and pressing upon the foreheads of the oxen, which is now common on the continent of Europe, and especially in France, is strongly condemned by Columella on grounds of economy as well as of humanity. (*De Re Rust.* ii.2.) He recommends that their heads should be left free, so that they may raise them aloft and thus make a handsomer appearance. All this was effected by the use either of the two collars shown in the upper figure of the woodcut, or of the excavations cut in the yoke, with the bands of leather, which are seen in the lower figure.

This [lower] figure also shows the method of tying the yoke to the pole by means of a leathern strap, which was lashed from the two opposite sides over the junction of the pole and yoke. These two parts were still more firmly connected by means of a pin which fitted a circular cavity in the middle of the yoke. Homer represents the leathern band as turned over the fastening thrice in each direction. But the fastening was sometimes much more complicated, especially in the case of the celebrated Gordian knot, which tied the yoke of a common cart, and consisted only of flexible twigs or bark, but in which the ends were so concealed by being inserted within the knot, that the only way of detaching the yoke was that which Alexander adopted.

Besides being variegated with precious materials and with carving, the yoke, especially among the Persians, was decorated with elevated plumes and figures. Of this an example is presented in a bas-relief from Persepolis, preserved in the British Museum. The chariot of Darius was remarkable for the golden statues of Belus and Ninus, about eighteen inches high, which were fixed to the yoke over the necks of the horses, a spread eagle, also wrought in gold, being placed between them.

IVDM RPLV
RPLV
EFAS
ALE

Libra

LIBRA, *dim.* LIBELLA (σταθμός), a balance, a pair of scales.

The principal parts of this instrument were, 1. The beam [Jugum, above], whence anything which is to be weighed is said to be "thrown under the beam."

2. The two scales, called in Greek τάλαντα (Hom. *Il.* viii.69, xii.433, xvi.659, xix.223, xxii.209; Aristoph. *Ranae,* 809) and πλάστιγγε (Aristoph. *Ranae,* 1425), and in Latin *lances* (Virg. *Aen.* xii.725; Pers. iv.10; Cic. *Acad.* iv.12). [Lanx.] Hence the verb ταλαντεύω is employed as equivalent to σταθμάω, and to the Latin *libro,* and is applied as descriptive of an eagle balancing his wings in the air. (Philostrat. Jun. *Imag.* 6; Welcker, *ad loc.*)

The beam was made without a tongue, being held by a ring or other appendage (*ligula, ῥῦμα*), fixed in the centre. (See the woodcut.)

Specimens of bronze balances may be seen in the British Museum and in other collections of antiquities, and also of the steel-yard [Statera], which was used for the same purposes as the libra.

In the works of ancient art, the balance is also introduced emblematically in a great variety of ways. The annexed woodcut is taken from a beautiful bronze patera, representing Mercury and Apollo engaged in exploring the fates of Achilles and Memnon, by weighing the attendant genius of the one against that of the other. (Winckelmann, *Mon. Ined.* 133; Millin, *Peintures de Vases Ant.* i. pl. 19, p. 39.)

A balance is often represented on the reverse of the Roman imperial coins; and to indicate more distinctly its signification, it is frequently held by a female in her right hand, while she supports a cornucopia in her left, the words *Aeqvitas Avgvsti* being inscribed on the margin, so as to denote the justice and impartiality with which the emperors dispensed their bounty.

The constellation Libra is placed in the Zodiac at the equinox, because it is the period of the year at which day and night are equally balanced. (Virg. *Georg.* i.208; Plin. *H.N.* xviii.69; Schol. *in Arat.* 89.)

The mason's or carpenter's level was called *libra* or *libella* (whence the English name), on account of its resemblance in many respects to a balance. (Varro, *de Re Rust.* i.6; Columella, iii.13; Plin. *H.N.* xxxvi.52.) Hence the verb *libro* meant to level as well as to weigh.

Machinae

MACHINAE (μηχαναί), and ORGANA (ὄργανα). The object of this article is to give a brief general account of those contrivances for the concentration and application of force, which are known by the names of *instruments, mechanical powers, machines, engines,* and so forth, as they were in use among the Greeks and Romans, especially in the time of Vitruvius, to whose tenth book the reader is referred for the details of the subject.

The general, but loose, definition which Vitruvius gives of a *machine* (x.1. § 1), is a wooden structure, having the virtue of moving very great weights. A *machina* differs from an *organon,* inasmuch as the former is more complex and produces greater effects of power than the latter: perhaps the distinction may be best expressed by translating the terms respectively *machine* or *engine* and *instrument.* Under the latter class, besides common *tools* and *simple instruments,* as the plough for example, Vitruvius appears to include the *simple mechanical powers,* which, however, when used in combination, as in the crane and other machines, become *machinae.* Thus Horace uses the word for the machines used to launch vessels (*Carm.* i.4.2), which appears to have been effected by the joint force of ropes and pulleys drawing the ship, and a screw pushing it forwards, aided by rollers (φάλαγγες) beneath it. The word *organon* was also used in its modern sense of a musical instrument.

The Greek writers, whom Vitruvius followed, divided machines into three classes, the (*genus*) *scanscorium* or ἀκροβατικόν (respecting which see Vitruvius and his commentators), the *spiritale* or πνευματικόν, and the *tractorium* or βαροῦλκον for moving heavy weights. The information which he gives us may perhaps, however, be exhibited better under another classification.

I. *Mechanical Engines.*

1. *The Simple Mechanical Powers* were known to the Greek mechanicians from a period earlier than can be assigned, and their theories were completely demonstrated by Archimedes. Vitruvius (x. 3. s.8) discourses of the two modes of raising heavy weights, by *rectilinear* and *circular* motion. He explains the action of the *lever (Ferreus vectis),* and its three different sorts, according to the position of the fulcrum, and some of its applications, as in the *steelyard (trutina, statera),* and the oars and rudder-oars of a ship; and alludes to the principle of *virtual velocities.*

[37]

The *inclined plane* is not spoken of by Vitruvius as a *machina*, but its properties as an aid in the elevation of weights are often referred to by him and other writers; and in early times it was, doubtless, the sole means by which the great blocks of stone in the upper parts of buildings could be raised to their places.

Under the head of circular motion, Vitruvius makes a passing allusion to the various forms of wheels and screws, *plaustra, rhedae, tympana, rotae, cochleae, scorpiones, balistae, prela.*

It is worth while, also, to notice the methods adopted by Chersiphron and his son Metagenes, the architects of the temple of Artemis at Ephesus, and by later architects, to convey large blocks of marble from the quarries, by supporting them in a cradle between wheels, or enclosing them in a cylindrical frame-work of wood (Vitruv. x.6. s.2); and also the account which Vitruvius gives of the mode of measuring the distance passed over by a carriage or a ship, by an instrument attached to the wheel of the former, or to a sort of paddle-wheel projecting from the side of the latter (c.9 s.14).

2. *Compound Mechanical Powers,* or *Machines for raising heavy weights (machinae tractoriae).* Of these Vitruvius describes three principal sorts, all of them consisting of a proper erect frame-work (either three beams, or one supported by ropes); from which hang *pullies,* the rope of which is worked either by a number of men, or by a windlass (*sucula*), or by a large drum (*tympanum,* ἀμφίφευσις, περιτρόχιον) moved as a tread-wheel, only from within. He describes the different sort of pullies, according to the number of *sheaves* (*orbiculi*) in each block (*trochlea* or *rechamus*), whence also the machine received special names, such as *trispastos,* when there were *three* sheaves, one in the lower block and two in the upper; and *pentaspastos,* when there were *five* sheaves, two in the lower block, and and three in the upper (x.2-5).

II. *Hydraulic Engines.*

1. *Conveyance and delivery of water through pipes and channels.* [Aquaeductus; Emissarium; Fistula; Fons.] The ancients well knew, and applied in practice, the hydrostatic law, that water enclosed in a bent pipe rises to the same level in both arms. It also appears, from the work of Frontinus, that they were acquainted with the law of hydraulics, that the quantity of water delivered by an orifice in a given time depends on the size of the orifice and on the height of the water in the reservoir; and also, that it is delivered faster through a short pipe than through a mere orifice of equal diameter.

[38]

2. *Machines for raising water.* The ancients did not know enough of the laws of atmospheric pressure to be acquainted with the common sucking pump; but they had a sort of forcing pump, which is described by Vitruvius (x.12), who ascribes the invention to Ctesibius.

For raising water a small height only they had the well-known screw of Archimedes, an instrument which, for this particular purpose, has never been surpassed [and which is described above: see "Cochlea"]. But their pumps were chiefly on the principle of those in which the water is lifted in buckets, placed either at the extremity of a lever, or on the rim of a wheel, or on a chain working between two wheels [see "Antlia," above]. (Vitruv. x.9.)

3. *Machines in which water is the moving power.* (Vitruv. x.10.) [See "Mola," below].

Mola

MOLA ($\mu\acute{\nu}\lambda o\varsigma$), a mill. All mills were anciently made of stone, the kind used being a volcanic trachyte or porous lava (*pyrites,* Plin. *H.N.* xxxvi.30; *silices,* Virg. *Moret.* 23-27; *pumiceas,* Ovid. *Fast.* vi.318), such as that which is now obtained for the same purpose at Mayen and other parts of the Eifel in Rhenish Prussia. This species of stone is admirably adapted for the purpose, because it is both hard and cavernous, so that, as it gradually wears away, it still presents an infinity of cutting surfaces.

Every mill consisted of two essential parts, the upper mill-stone, which was moveable (*catillus,* $\acute{\nu}\nu o\varsigma$, $\tau\grave{o}$ $\grave{\epsilon}\pi\iota\mu\acute{\nu}\lambda\iota o\nu$, *Deut.* xxiv.6), and the lower, which was fixed and by much the larger of the two. Hence a mill is sometimes called *molae* in the plural. The mills mentioned by ancient authors are the following:—

I. The hand-mill, or quern, called *mola manuaria, versatilis,* or *trusatilis.* (Plin. *H.N.* xxxvi.29; Gell. iii.3; Cato, *de Re Rust.* 10.)

The islanders of the Archipelago use in the present day a mill, which consists of two flat round stones about two feet in diameter. The upper stone is turned by a handle ($\kappa\acute{\omega}\pi\eta$, Schol. *in Theocrit.* iv.58) inserted at one side, and has a hole in the middle into which the corn is poured. By the process of grinding the corn makes its way from the centre, and is poured out in the state of flour at the rim. (Tournefort, *Voyage, Lett.* 9.) The description of this machine exactly agrees with that of the Scottish quern, formerly an indispensable part of domestic furniture. (Pennant, *Tour in Scotland,* 1769, p.231; and 1772, p.328.)

There can be no doubt that this is the flour-mill in its most ancient form. In a very improved state it has been discovered at Pompeii. The annexed woodcut shows two which were found standing in the ruins of a bakehouse. In the left-hand figure the lower millstone only is shown. The most essential part of it is the cone, which is surmounted by a projection containing originally a strong iron pivot. The upper millstone, seen in its place on the right hand of the woodcut, approaches the form of an hour-glass, consisting of two hollow cones, jointed together at the apex, and provided at this point with a socket, by which the upper stone was suspended upon the iron pivot, at the same time touching on all sides the lower stone, and with which it was intended to revolve. The upper stone was surrounded at its narrowest part with a strong band of iron; and two bars of wood were inserted into square holes, one of which appears in the figure, and were used to turn the upper stone. The uppermost of the two hollow cones served the purpose of a hopper. The corn with which it was filled, gradually fell through the neck of the upper stone upon the summit of the lower, and, as it proceeded down the cone, was ground into flour by the friction of the two rough surfaces, and fell on all sides of the base of the cone into a channel formed for its reception. The mill here represented is five or six feet high.

The hand-mills were worked among the Greeks and Romans by slaves. Their pistrinum was consequently proverbial as a place of painful and degrading labour; and this toil was imposed principally on women. (Hom. *Od.* vii.104; Exod. xi.5; Matt. xxiv.41.)

In every large establishment the hand-mills were numerous in proportion to the extent of the family. Thus in the palace of Ulysses there were twelve, each turned by a separate female, who was obliged to grind every day the fixed quantity of corn before she was permitted to cease from her labour. (*Od.* xx.105-119; compare Cato, *de Re Rust.* 56.)

II. The cattle-mill, *mola asinaria* (Cato, *de Re Rust.* 10; Matt. xviii.6) in which human labour was supplied by the use of an ass or some other animal. (Ovid, *Fast.* vi.318.) The animal devoted to this labour was blind-folded. (Apul. *Met.* ix.) The mill did not differ in its construction from the larger kinds of hand-mill.

III. The water-mill (*mola aquaria, ὑδραλέτης*). The first water-mill, of which any record is preserved, was connected with the palace of Mithridates in Pontus. (Strabo, xii.3. § 30.) That water-mills were used at Rome is manifest from the description of them by Vitruvius (x.5. ed. Schneider). A cogged wheel, attached to the axis of the water-wheel, turned another which was attached to the axis of the upper mill-stone: the corn to be ground fell between the stones out of a hopper (*infundibulum*), which was fixed above them. (See also Brunck, *Anal.* ii.119; Pallad. *de Re Rust.* i.42.) Ausonius, as quoted below, mentions their existence on the Ruwer near Treves; and Venantius Fortunatus, describing a castle built in the sixth century on the banks of the Moselle, makes distinct mention of a tail-race, by which "the tortuous stream is conducted in a straight channel." (*Poem.* iii.10.)

[42]

IV. The floating-mill. When Rome was besieged by the Goths, A.D. 536, and when the stoppage of the aqueducts rendered it impossible to use the public corn-mills (οἱ τῆς πόλεως μύλωνες) in the Janiculum, so that the citizens were in danger of starvation, Belisarius supplied their place by erecting floating-mills upon the Tiber. Two boats being moored at the distance of two feet from each other, a water-wheel, suspended on its axis between them, was turned by the force of the stream, and put in motion the stones for grinding the corn, by which the lives of the besieged were preserved. (Procop. *de Bello Gothico*, i.15.)

V. The saw-mill. Ausonius mentions mills situated on some of the streams falling into the Moselle, and used for cutting marble into slabs. (*Mosella*, 362, 363.)

VI. The pepper-mill. A mill for grinding pepper, made of boxwood, is mentioned by Petronius (*molea buxea piper trivit, Sat.* 74).

Fig. 1

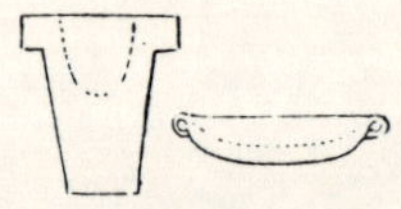

Fig. 2

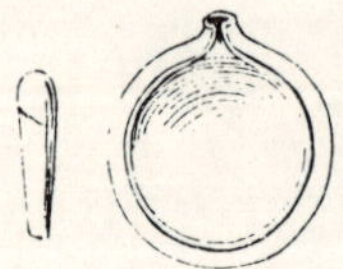

Fig. 3

Mortarium

MORTARIUM, also called PILA and PILUM (Plin. *H.N.* xviii.3; xxxiii.26), (ὄλμος: ἴγδη, Schol. *in Hes. Op. et Dies,* 421; ἴγδις, apparently from the root of *icere,* to strike), a mortar.

Before the invention of mills [see "Mola," above] corn was pounded and rubbed in mortars (*pistum*), and hence the place for making bread, or the bakehouse, was called *pistrinum.* (Serv. *in Virg. Aen.* i.179.) Also long after the introduction of mills this was an indispensable article of domestic furniture. (Plaut. *Aul.* i.2.17; Cato, *de Re Rust.* 74-76; Colum. *de Re Rust.* xii.55.)

Hesiod (*l.c.*), enumerating the wooden utensils necessary to a farmer, directs him to cut a mortar three feet, and a pestle (ὕπερον, κοπάνον, *pistillum*) three cubits long. Both of these were evidently to be made from straight portions of the trunks or branches of trees, and the thicker and shorter of them was to be hollowed. They might then be used in the manner represented in a painting on the tomb of Rameses III of Egypt at Thebes (see woodcut, figure no. 1, taken from Wilkinson, vol. ii. p. 383); for there is no reason to doubt that the Egyptians and the Greeks fashioned and used their mortars in the same manner. (See also Wilkinson, vol. iii. p. 181, showing three stone mortars with metal pestles.) In these paintings we may observe the thickening of the pestle at both ends, and that two men pound in one mortar, raising their pestles alternately as is still the practice in Egypt.

Pliny (*H.N.* xxxvi.43) mentions the various kinds of stone selected for making mortars, according to the purposes to which they were intended to serve. Those used in pharmacy were sometimes made, as he says, "of Egyptian alabaster." The second woodcut shows the forms of two preserved in the Egyptian collection of the British Museum, which exactly answer to this description, being made of that material. They do not exceed three inches in height: the dotted lines mark the cavity within each.

The third woodcut shows a mortar and pestle, made of baked white clay, which were discovered, A.D. 1831, among numerous specimens of Roman pottery in making the northern approaches to London-bridge (*Archaeologia,* vol. xxiv. p. 199, plate 44.)

Besides the uses already mentioned, the mortar was employed in pounding charcoal, rubbing it with glue, in order to make black paint (*atramentum,* Vitruv. vii.10. ed. Schneider); in making plaster for the walls of apartments (Plin. *H.N.* xxxvi.55); in mixing spices and fragrant herbs and flowers for the use of the kitchen (Athen. ix.70; Brunck, *Anal.* iii.51); and in metallurgy, as in triturating cinnabar to obtain mercury from it by sublimation. (Plin. *H.N.* xxxiii.41, xxxiv.22.)

Pegma

PEGMA (πῆγμα), a pageant, *i.e.* an edifice of wood, consisting of two or more stages (*tabulata*), which were raised or depressed at pleasure by means of balance-weights (*ponderibus reductis,* Claudian, *de Mallii Theod. Cons.* 323-328; Sen. *Epist.* 89).

These great machines were used in the Roman amphitheatres (Juv. iv.121; Mart. i.2.2; Sueton. *Claud.* 34), the gladiators who fought upon them being called *pegmares.* (*Calig.* 26.) They were supported upon wheels so as to be drawn into the circus, glittering with silver and a profusion of wealth. (Plin. *H.N.* xxxiii.3 s.16.)

At other times they exhibited a magnificent though dangerous (Vopisc. *Carin.* 15) display of fireworks. (Claudian, *l.c.*) Accidents sometimes happened to the musicians and other performers who were carried upon them. (Phaedr. v.7.7.)

The pegmata mentioned by Cicero (*ad Att.* iv.8) may have been movable book-cases.

Phalangae

PHALANGAE or PALANGAE (φάλαγγες), any long cylindrical pieces of wood, such as trunks or branches of trees (Herod. iii.97; Plin. *H.N.* xii.4 s.8), truncheons (Plin. *H.N.* vii.56. s.57), and poles used to carry burthens. The carriers who used these poles were called *phalangarii* (*Gloss. Ant. s.v.*), and also *hexaphori, tetraphori,* &c., according as they worked in parties of six, four, or two persons.

The word was especially used to signify rollers placed under ships to move them on dry land, so as to draw them upon shore or into the water (δουρατέοι κυλίνδροι, Brunck, *Anal.* iii.89; Apoll. Rhod. i.375-389). This was effected either by making use of the oars as levers, and at the same time fastening to the stern of the ship cables with a noose (μηρίνθος), against which the sailors pressed with their breasts, as we see in our canal navigation (Orph. *Argon.* 239-249, 270-273), or by the use of machines. (Hor. *Carm.* i.4.2.)

Rollers were employed in the same manner to move military engines (Caesar, *Bell. Civ.* ii.10).

[48]

RETIS, RETE

RETIS and RETE; *dim.* RETICULUM (δίκτυον), a net. Nets were made most commonly of flax from Egypt, Colchis, the vicinity of the Cinyps in North Africa, and some other places. Occasionally they were of hemp. (Varro, *de Re Rust.* iii.5.) They are sometimes called *lina* (λίνα) on account of the material of which they consisted. The meshes were great or small according to the purposes intended; and these purposes were very various.

By far the most important application of net-work was to the three kindred arts of fowling, hunting, and fishing: and besides the general terms used alike in reference to all these employments, there are special terms to be explained under each of these heads.

I. In fowling the use of nets was comparatively limited (Aristoph. *Av.* 528); nevertheless thrushes were caught in them; and doves or pigeons with their limbs tied up or fastened to the ground, or with their eyes covered or put out, were confined in a net, in order that their cries might allure others into the snare. The ancient Egyptians, as we learn from the paintings in their tombs, caught birds in clap-nets. (Wilkinson, *Man and Cust.* vol. iii. pp.35-38, 45.)

II. In hunting it was usual to extend nets in a curved line of considerable length, so as in part to surround a space into which the beasts of chase, such as the hare, the boar, the deer, the lion, and the bear, were driven through the opening left on one side. This range of nets was flanked by cords, to which feathers dyed scarlet and of other bright colours were tied, so as to flare and flutter in the wind. The hunters then sallied forth with their dogs, dislodged the animals from their coverts, and by shouts and barking drove them first within the *formido,* as the apparatus of string and feathers was called, and then, as they were scared with this appearance, within the circuit of the nets. Splendid descriptions of this scene are given in some of the following passages, all of which allude to the spacious enclosure of net-work. (Virg. *Georg.* iii.411-413, *Aen.* iv.121, 151-159, x.707-715; Ovid. *Epist.* iv.41,42 v.19,20; Oppian, *Cyn.* iv.120-123; Eurip. *Bacchae,* 821-832.)

The accompanying woodcuts are taken from two bas-reliefs in the collection of ancient marbles at Ince-Blundell in Lancashire. In the uppermost figure three servants with staves carry on their shoulders a large net, which is intended to be set up as already described. The foremost servant holds by a leash a dog, which is eager to pursue the game.

In the middle figure the net is set up. At each end of it stands a watchman holding a staff. Being intended to take such large quadrupeds as boars and deer (which are seen within it), the meshes are very wide. The net is supported by three stakes. To dispose the nets in this manner was called *retia ponere* (Virg. *Georg.* i.307), or *retia tendere* (Ovid. *Art. Amat.* i.45). Comparing it with the stature of the attendants, we perceive the net to be between five and six feet high. The upper border of the net consists of a strong rope, which was called σαρδών. (Xen. *de Venat.* vi.9.)

The figures in the third woodcut represent two men carrying the net home after the chase; the stakes for supporting it, two of which they hold in their hands, are forked at the top, as is expressed by the terms *ancones* and *vari.*

Besides the nets used to inclose woods and coverts or other large tracts of country two additional kinds are mentioned by those authors who treat on hunting. These were placed at intervals in the same circuit with the large hunting-net or haye. The road-net was much less than the others, and was placed across roads and narrow openings between bushes. The purse-net or tunnel-net (*cassis,* ἄρκυς) was made with a bag, intended to receive the animal when chased towards the extremity of the inclosure. Within this bag, if we may so call it, were placed branches of trees, to keep it expanded and to decoy the animals by making it invisible.

III. Fishing-nets (ἁλιευτικὰ δίκτυα, Diod. Sic. xvii.43, p. 193, Wess.) were of six different kinds. Of these by far the most common were the casting-net and the drag-net.

The drag-net (sean) which, as now used both by the Arabians and by fishermen in Cornwall, is sometimes half a mile long, and was probably of equal dimensions among the ancients, for they speak of it as nearly taking in the compass of a whole bay. (Hom. *Od.* xxii. 384-387; Alciphron, i.17,18.) The use of corks to support the top, and of leads to keep down the bottom, is frequently mentioned by ancient writers, and is clearly exhibited in some of the paintings in Egyptian tombs. Leads, and pieces of wood serving as floats instead of corks, still remain on a drag-net which is preserved in the fine collection of Egyptian antiquities at Berlin.

The casting-net, being always pear-shaped or conical, was suited to use as a mosquito-net, to be expanded over beds and couches to keep away gnats and other flying insects. The use of them is still common in Italy, Greece, and other countries surrounding the Mediterranean. The Greek word for a casting-net (ἀμφίβληστρον) is used metaphorically to denote some certain method of destruction, and is more particularly applied to the large shawl in which Clytemnestra enveloped her husband in order to murder him. (Aeschyl. *Agam.* 1085, 1346, 1353, *Choeph.* 485, *Eumen.* 112.)

Scalae

SCALAE (κλῖμαξ), a ladder. The general construction and use of ladders was the same among the ancients as in modern times, and therefore requires no explanation, with the exception of those used in besieging a fortified place and in making an assault upon it.

The ladders were erected against the walls, and the besiegers ascended them under showers of darts and stones thrown upon them by the besieged. Some of these ladders were formed like our common ones; others consisted of several parts which might be put together so as to form one large ladder, and were taken apart when they were not used.

Sometimes also they were made of ropes or leather with large iron hooks at the top, by which they were fastened to the walls to be ascended. The ladders made wholly of leather consisted of tubes sowed up air-tight, and when they were wanted, these tubes were filled with air. There was also a ladder constructed in such a manner that it might be erected with a man standing on the top, whose object was to observe what was going on in the besieged town. Others again were provided at the top with a small bridge, which might be let down upon the wall. In ships small ladders or steps were likewise used for the purpose of ascending into or descending from them.

In the houses of the Romans the name Scalae was applied to the stairs or staircase, leading from the lower to the upper parts of a house. The steps were either of wood or stone, and, as in modern times, fixed on one side in the wall. It appears that the staircases in Roman houses were as dark as those of old houses today, for it is very often mentioned that a person concealed himself *in scalis* or *in scalarum tenebris* (Cic. *pro Mil.* 15, *Philip.* ii.9; Horat. *Epist.* ii.2.15). The Roman houses had two kinds of staircases: the common scalae, open on one side, and the Greek scalae, which were closed on both sides.

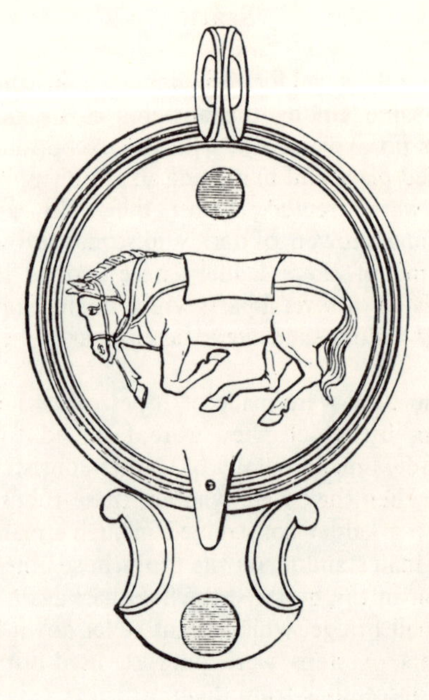

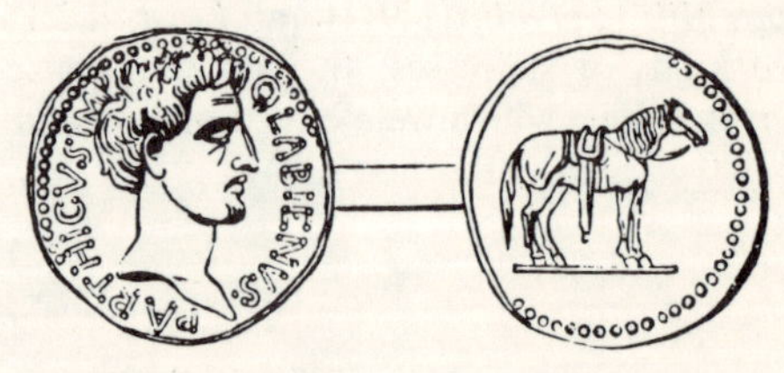

PARTHICVS
Q·LABIENVS·

Tela

TELA (ἰστός), a loom. Although weaving was amongst the Greeks and Romans a distinct trade carried on by a separate class of persons, who more particularly supplied the inhabitants of the towns with the productions of their skill, yet every considerable domestic establishment, especially in the country, contained a loom together with the whole apparatus necessary for the working of wool. These occupations were all supposed to be carried on under the protection of Minerva, specially denominated Ἐργάνη, who was always regarded in this character as the friend and patroness of industry, sobriety, and female decorum.

When the farm or the palace was sufficiently large, a portion of it called the ἰστών or *textrinum,* was devoted to this purpose. The work was there carried on principally by female slaves under the superintendence of the mistress of the house, who herself also together with her daughters took part in the labour, both by instructing beginners and by finishing the more tasteful and ornamental parts.

But although weaving was employed in providing the ordinary articles of clothing among the Greeks and Romans from the earliest times, yet as an inventive and decorative art, subservient to luxury and refinement, it was almost entirely Oriental. Persia, Babylonia, Egypt, Phoenicia, Phrygia, and Lydia are all celebrated for the wonderful skill and magnificence displayed in the manufacture of scarfs, shawls, carpets and tapestry.

Among the peculiarities of Egyptian manners Herodotus (ii.35) mentions that weaving was in that country the employment of the male sex. This custom still continues among some Arab and negro tribes. Throughout Europe, on the other hand, weaving was in the earliest ages the task of women only. The matron, assisted by her daughters, wove clothing for the husband and the sons.

Besides the shawls which were frequently given to the temples by private persons, or obtained by commerce with foreign nations, companies or colleges of females were attached to the more opulent temples for the purpose of furnishing a regular supply. Thus the sixteen women, who lived together in a building destined to their use at Olympia, wove a new shawl every five years to be displayed at the games which were then celebrated in honour of Hera, and to be preserved in her temple. A similar college at Sparta was devoted to the purpose of weaving a tunic every year for the sitting statue of the

Fig. 1

Amyclean Apollo, which was thirty cubits high. At Athens the company of virgins called ἐργαστῖναι or ἐργάναι, who were partly of Asiatic extraction, wove the shawl which was carried in the Panathenaic procession and which represented the battle between the gods and the giants. A similar occupation was assigned to young females of the highest rank at Argos.

In the fourth century the task of weaving began to be transferred in Europe from women to the other sex, a change which St. Chrysostom deplores as a sign of prevailing sloth and effeminacy.

Every thing woven consists of two essential parts, the warp and the woof. Plato mentions one of the most important differences between the two: namely that the threads of the warp are strong and firm in consequence of being more twisted in spinning, whilst those of the woof are comparatively strong and yielding (Plato, *Leg.* v. p. 386, ed. Bekker). This is in fact the difference which in the modern silk manufacture distinguishes *organzine* from *tram,* and in the cotton manufacture *twist* from *weft.*

The very first operation in weaving was to set up the loom, with the web or cloth hanging from the transverse beam ["Jugum," above], as is shown in the picture of Circe's loom, which is contained in the very ancient illuminated MS. of Virgil's Aeneid preserved at Rome in the Vatican Library (see figure 1). Although the upright loom here exhibited was in common use, and employed for all ordinary purposes, the practice, now generally adopted, of placing the warp in a horizontal position was occasionally resorted to in ancient times; for the upright loom, the management of which required the female to stand and move about, is opposed to another kind at which she sat.

We observe in the preceding woodcut (figure 1) about the middle of the apparatus a transverse rod passing through the warp. A straight cane was well adapted to be so used, and its application is clearly expressed by Ovid (*Met.* vi.55). In plain weaving it was inserted between the threads of the warp so as to divide them into two portions, the threads on one side of the rod alternating with those on the other side throughout the whole breadth of the warp.

[57]

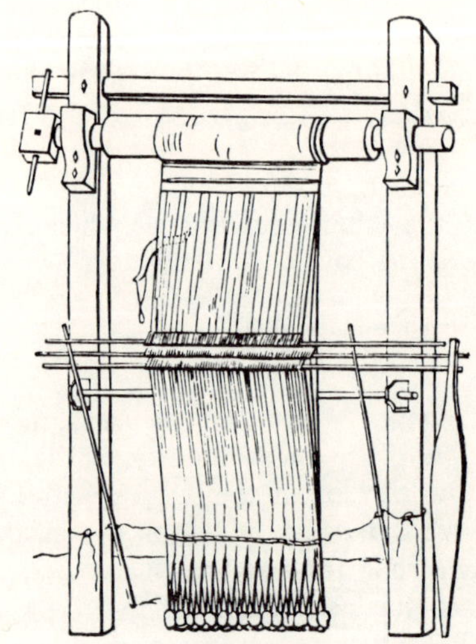

Fig. 2

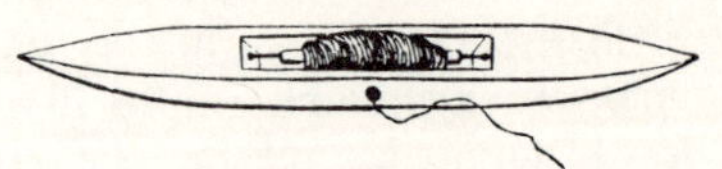

Fig. 3

Whilst the improvements in machinery have to a great extent superseded the use of the upright loom in all other parts of Europe, it remains almost in its primitive state in Iceland. The woodcut (figure 2) is reduced from an engraving of the Icelandic loom in Olaf Olafsen's *Economic Tour* in that island, published in Danish at Copenhagen, 1780. We observe underneath the jugum a roller which is turned by a handle, and on which the web is wound as the work advances. The threads of the warp, besides being separated by a transverse rod or plank, are divided into thirty or forty parcels, to each of which a stone is suspended for the purpose of keeping the warp in a perpendicular position and allowing the necessary play to the strokes of the spatha, which is drawn at the side of the loom. The mystical ode written about the eleventh century of our era, with which Gray has made us familiar in his translation, and which describes the loom of "the Fatal Sisters," represents warriors' skulls as supplying the place of these round stones. The knotted bundles of threads, to which the stones were attached, often remained after the web was finished in the form of a fringe.

Whilst the comparatively coarse, strong, and much-twisted thread designed for the warp was thus arranged in parallel lines, the woof remained upon the spindle, forming a *spool, bobbin,* or *pen.* This was either conveyed through the warp without any additional contrivance, as is still the case in Iceland, or it was made to revolve in a shuttle. This was made of box brought from the shores of the Euxine (Black Sea), and was pointed at its extremities, that it might easily force its way through the warp. The annexed woodcut (figure 3) shows the form in which it is still used in some retired parts of England for common domestic purposes, and which may be regarded as a form of great antiquity. An oblong cavity is seen in its upper surface, which holds the bobbin. A small stick, like a wire, extends through the length of this cavity and enters its two extremities so as to turn freely. The small stick passes through a hollow cane, called a *quill,* which is surrounded by the woof. This is drawn through a round hole in the front of the shuttle, and, whenever the shuttle is thrown, the bobbin revolves and delivers the woof through this hole.

All that is effected by the shuttle is the conveyance of the woof across the warp. To keep every thread of the woof in its proper place it is necessary that the threads of the warp should be decussated [decussate — to cross or cut in the form of an X; to intersect]. This was done by the leashes, a leash being a thread having at one end a loop through which a thread of the warp was passed, the other end being fastened to a straight rod called *Liciatorium.* The warp, having been divided by the arundo (the transverse rod, or cane, in figure 1), into two sets of threads, all those of the same set were passed through the loops of the corresponding sets of leashes, and all these leashes were fastened at their other end to the same wooden rod. At least one set of leashes was necessary to decussate the warp, even in the plainest and simplest weaving. The number of sets was increased according to the complexity of the pattern, that is, according as the number was two, three, or more.

The process of annexing the leashes to the warp occupied two women at the same time, one of whom took in regular succession each separate thread of the warp and handed it over to the other. The other woman, as she received each thread, passed it through the loop in proper order.

Supposing the warp to have been thus adjusted, and the pen or the shuttle to have been carried through it, it was then decussated by drawing forwards the proper rod, so as to carry one set of the threads of the warp across the rest, after which the woof was shot back again, and by the continual repetition of this process the warp and woof were interlaced. In the figure (2) of the Icelandic loom we observe two staves, which are occasionally used to fix the rods in such a position as is most convenient to assist the weaver in drawing her woof across her warp.

After the woof had been conveyed by the shuttle through the warp, it was driven sometimes downwards, as is represented in figure 1, but more commonly upwards, as in figure 2. Two different instruments were used in this part of the process. The simplest and probably the most ancient was in the form of a large wooden sword. This instrument is still used in Iceland exactly as it was in ancient times, and a figure of it copied from Olafsen, is given in figure 2.

The spatha (figure 2) was, however, in a great degree superseded by the comb, the teeth of which were inserted between the threads of the warp, and thus made by a forcible impulse to drive the threads of the woof close together. It is probable that the teeth were sometimes made of metal and accommodated to the intended purpose by being curved, as is still the case in the combs which are used in the same manner by the Hindoos.

It remains to describe the methods of producing the varieties of cloth, and more especially of adding to its value by making it either warmer and softer, or more rich and ornamental. If the object was to produce a checked pattern, or to weave what we should call a Scotch

[60]

plaid, the threads of the warp were arranged alternately black and white, or of different colours in a certain series according to the pattern which was to be exhibited. On the other hand, a striped pattern was produced by using a warp of one colour only, but changing at regular intervals the colour of the woof. Checked and striped goods were, no doubt, in the first instance, produced by combining the natural varieties of wool — white, black, brown, &c.

The woof also was the medium through which almost every other diversity of appearance and quality was effected. The warp was generally more twisted and consequently stronger and firmer than the woof. The consequence was that after the piece was woven the fuller drew out its nap by carding, so as to make it like a soft blanket. When the intention was to guard against the cold, the warp was diminished and the woof or nap made more abundant in proportion. On the other hand a woof of finely twisted thread produced a thin kind of cloth which resembled our buntine.

Where any kind of cloth was enriched by the admixture of different materials, the richer and more beautiful substance always formed part of the woof. Thus the *vestis subserica,* or *tramoserica,* had the tram of silk. In other cases it was of gold; of wool dyed with Tyrian purple; or of beavers'-wool.

Besides the variety of materials constituting the woof, an endless diversity was effected by the manner of inserting them into the warp. The terms *bilix* and δίμιτος probably denoted what we call *dimity* or *tweeled* cloth, and the Germans *zwillich.* The poets apply *trilix*, which in German has become *drillich,* to a kind of armour, perhaps chain-mail, no doubt resembling the pattern of cloth, which was denoted by the same term. In figure 2 of the Icelandic loom the three rods with their leashes indicate the arrangement necessary for this texture. All kinds of damask were produced by a very complicated apparatus of the same kind and were called *Polymita.*

As far as we can form a judgment from the language and descriptions of ancient authors, the productions of the loom appear to have fallen in ancient times very little, if at all, below the beauty and variety of the damasks, shawls, and tapestry of the present age, and to have vied with the works of the most celebrated painters, representing first mythological, and afterwards scriptural subjects.

[61]

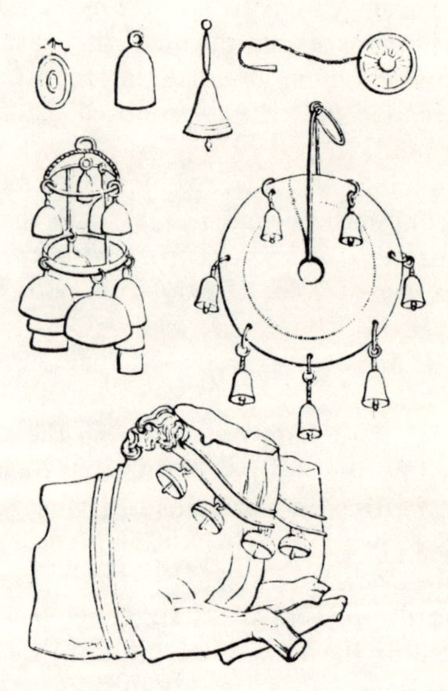

Tintinnabulum

TINTINNABULUM (κώδων), a bell. Bells were used for a great variety of purposes among the Greeks and Romans, which it is unnecessary to particularize here. One use, however, was for the purpose of keeping watch and ward in the fortified cities of Greece, and deserves mention.

A guard (φύλαξ) being stationed in every tower, a περίπολος (a youth serving a kind of apprenticeship in arms) walked to and fro on the portion of the wall between two towers. It was his duty to carry the bell, which he received from the guard at one tower, to deliver it to the guard at the next tower, and then to return, so that the bell by passing from hand to hand made the circuit of the city. By this arrangement it was discovered if any guard was absent from his post, or did not answer to the bell in consequence of being asleep.

The forms of bells were various in proportion to the multiplicity of their applications. In the Museum at Naples are some of the form which we call bell-shaped; others are more like a Chinese gong. The bell, figure 1, is a simple disk of bell-metal; it is represented in a painting as hanging from the branch of a tree. Figure 2 represents a bell of the same form, but with a circular hole in the centre, and a clapper attached to it by a chain. This is in the Museum at Naples, as well as the bell, figure 3, which in form is exactly like those still commonly used in Italy to be attached to the necks of sheep, goats, and oxen. Figure 4 is represented on one of Sir W. Hamilton's vases (i.43) as carried by a man in the garb of Pan, and probably for the purpose of lustration. (Theocrit. ii.36; Schol. *in loc.*) Figure 5 is a bell, or rather a collection of twelve bells suspended in a frame, which is preserved in the Antiquarium at Munich. This jingling instrument, as well as that represented by figure 6 (from Bartoli, *Luc. Sep.* ii.23), may have been used at sacrifices, in Bacchanalian processions, or for lustration. Figure 7 is a fragment of ancient sculpture, representing the manner in which bells were attached to the collars of chariot-horses.

[64]

Torculum

TORCULUM or TORCULAR (ληνός), a press for making wine and oil. When the grapes were ripe, the bunches were gathered, any which remained unripe or had become dry or rotten were carefully removed, and the rest carried from the vineyard in deep baskets to be poured into a shallow vat. In this they were immediately trodden by men, who had the lower part of their bodies naked, except that they wore drawers. At least two persons usually trod the grapes together. To "tread the wine-press alone" indicated desolation and distress.

The Egyptian paintings (Wilkinson, *Man. and Cust.* vol. ii. pp. 152-157) exhibit as many as seven treading in the same vat, and supporting themselves by taking hold of ropes or poles placed above their heads. From the size of the Greek and Roman vats there can be no doubt that the company of treaders was often still more numerous. To prevent confusion and to animate them in their labour they moved in time or danced, as is seen in the ancient mosaics of the church of St. Constantia at Rome, sometimes also leaning upon one another, as can be seen in the annexed woodcut. Sometimes a person by the side of the vat played instrumental music for the purpose of aiding and regulating the movements of those in it. Besides the music they were cheered with a song.

After the grapes had been trodden sufficiently, they were subjected to the more powerful pressure of a thick and heavy beam for the purpose of obtaining all the juice yet remaining in them. A press with a screw was sometimes used instead of a beam. A strainer or colander was then employed to clear the must from solid particles, as it flowed from the vat.

The woodcut shows the apertures at the bottom of the vat, by which the must was discharged, and the method of receiving it, when the vat was small, in wide-mouthed jars, which when full were carried away to be emptied into casks. When the vineyard was extensive and the vat large in proportion, the must flowed into another vat of corresponding size, which was sunk below the level of the ground.

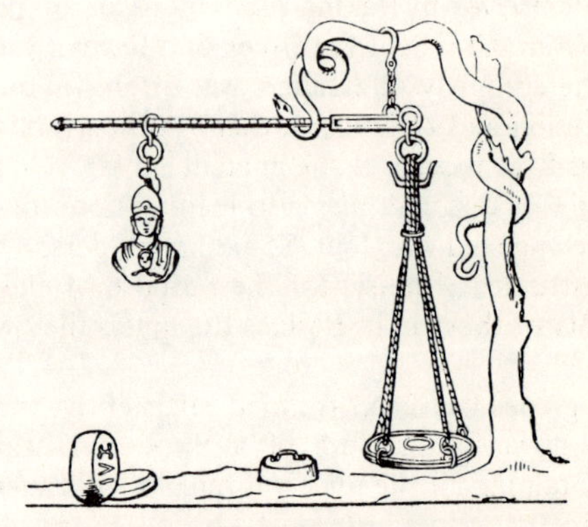

Trutina

TRUTINA (τρυτάνη), a general term including both Libra (which see, above), a balance, and *statera,* a steelyard. Payments were originally made by weighing, not by counting. Hence a balance (*trutina*) was preserved in the temple of Saturn at Rome.

The balance was much more ancient than the steelyard, which according to Isidore of Seville (*Orig.* xvi.24) was invented in Campania, and therefore called by way of distinction *Trutina Campana.* Consistently with this remark, steelyards have been found in great numbers among the ruins of Herculaneum and Pompeii. The construction of some of them is more elaborate and complicated than that of modern steelyards, and they are in some cases much ornamented.

The annexed woodcut represents a remarkably beautiful statera which is preserved in the Museum of the Capitol at Rome. Its support is the trunk of a tree, round which a serpent is entwined. The equipoise is a head of Minerva. Three other weights lie on the base of the stand, designed to be hung upon the hook when occasion required. (*Mus. Capit.* vol. ii. p. 213.)

Vitruvius (x.3. s.8. § 4) explains the principle of the steelyard, and mentions the following constituent parts of it: the scale (*lancula*) depending from the head (*caput*), near which is the point of revolution (*centrum*) and the handle (*ansa*). On the other side of the centre from the scale is the beam (*scapus*) with the weight or equipoise (*aequipondium*), which is made to move along the points (*per puncta*) expressing the weights of the different objects that are put into the scale.